16 Medicinal Plants

To Keep In Your House
Bilingual Edition English Germany
Standar Version

by

Jannah Firdaus Mediapro

2020

16 Medicinal Plants To Keep in Your House Bilingual Edition English Germany Standar Version

Copyright © 2020

Jannah Firdaus Mediapro

All rights reserved

16 Medicinal Plants To Keep in Your House Bilingual Edition English Germany Standar Version

Prologue

16 Medicinal Plants to Keep in Your House Bilingual Edition In English and Germany Languange Standar Version.

Medicinal plants, also called medicinal herbs, have been discovered and used in traditional medicine practices since prehistoric times. Plants synthesise hundreds of chemical compounds for functions including defence against insects, fungi, diseases, and herbivorous mammals. Numerous phytochemicals with potential or established biological activity have been identified.

The earliest historical records of herbs are found from the Sumerian civilisation, where hundreds of medicinal plants including opium are listed on clay tablets. The Ebers Papyrus from ancient Egypt, c. 1550 BC, describes over 850 plant medicines. The Greek physician Dioscorides, who worked in the Roman army, documented over 1000 recipes for medicines using over 600 medicinal plants in De materia medica, c. 60 AD; this formed the basis of pharmacopoeias for some 1500 years. Drug research makes use of ethnobotany to search for pharmacologically active substances in nature, and has in this way discovered hundreds of useful compounds. These include the common drugs aspirin, digoxin, quinine, and opium. The compounds found in plants are of many kinds, but most are in four major biochemical classes: alkaloids, glycosides, polyphenols, and terpenes.

16 Medicinal Plants To Keep in Your House Bilingual Edition English Germany Standar Version

Medicinal plants are useful to keep on hand to treat common ailments. You can reach for certain medical plants to relieve headaches, tummy trouble and even irritation from bug bites. Plants can be consumed in teas, used as garnish, applied topically as essential oil or consumed as a pill.

It's important to remember that you should always double check with your doctor before consuming or using anything new for your body. If you choose to grow some of these plants, remember to take proper care according to the plant's care guidelines and refrain from using any pesticides or other harmful chemicals on your plants. You don't want any of those chemicals in or on your body!

To help you decide what plants are best for you, we rounded up our top medicinal plants, their notable health benefits and how to use them.

Medicinal Plants to Keep in Your House English Edition

Medicinal plants are useful to keep on hand to treat common ailments. You can reach for certain medical plants to relieve headaches, tummy trouble and even irritation from bug bites. Plants can be consumed in teas, used as garnish, applied topically as essential oil or consumed as a pill.

It's important to remember that you should always double check with your doctor before consuming or using anything new for your body. If you choose to grow some of these plants, remember to take proper care according to the plant's care guidelines and refrain from using any pesticides or other harmful chemicals on your plants. You don't want any of those chemicals in or on your body!

To help you decide what plants are best for you, we rounded up our top medicinal plants, their notable health benefits and how to use them.

01. BASIL

1. Basil

Basil (*ocimum basilicum*) is a common herb used to garnish salads, pasta and many other meals to add delicious flavor. Thanks to the vitamins and minerals in basil, such as vitamin K and iron, this herb is helpful for combating common ailments. For example, the manganese in basil helps metabolize different compounds in your body. Holy basil, commonly referred to as tulsi, is a specific species of basil that originates from India. It's considered a sacred plant that is used in teas, ointments and more, to help treat a variety of ailments like fevers and diabetes. This species has a much stronger taste than common basil!

Basil health benefits:

- Reduces stress
- Strong antibacterial properties
- Rich source of antioxidants

- Prevents some harmful effects of aging
- Reduces inflammation and swelling
- Strengthens bones and liver
- Boosts immunity
- Boosts metabolism
- Improves digestion

Common uses:

- Sprinkle as a garnish for dishes
- Include as ingredient for smoothies

02. CATNIP

2. Catnip

Catnip (*nepeta cataria*) is a fun plant for cats. Most cats are attracted to the plant and will roll around near it since its aroma acts as a stimulant. These medicinal plants also act as a sedative for cats if consumed. For humans, on the other hand, it is normally used as a stress reliever, sleep aid and a solution for skin issues. The majority of its health benefits come from the presence of nepetalactone, thymol and other compounds that make this plant great for you and your furry friend.

Catnip health benefits:

- Repels bugs and relieves irritation from bug bites
- Calms restlessness, anxiety and stress
- Relieves stomach discomfort
- Accelerates recovery from colds and fevers

Common uses:

- Brew leaves for a tea
- Dry leaves and burn to release aroma
- Apply essential oils or leaves topically

03. CAYENNE PEPPER

3. Cayenne Pepper

Cayenne pepper (*capsicum annuum*) adds a spicy kick to any meal or drink and is a popular detoxifier for many people. Capsaicin is the compound responsible for cayenne's spicy nature, but it's also responsible for some of its health benefits. Some of these benefits include pain relief and lower cholesterol.

Cayenne pepper health benefits:
- Detoxifies the body
- Boosts metabolism
- Eases an upset stomach and helps digestion
- Improves circulation
- Relieves pain

Common uses:

- Add to sauces, spice mixes, dressing and other dishes
- Consume as a pill

04. CHAMOMILE

4. Chamomile

Chamomile (*matricaria chamomilla*) has a high concentration of antioxidants that make it a great plant for relieving a variety of ailments. Chamomile is commonly consumed as a tea and you can make your own at home by brewing dried chamomile flowers (just make sure the flowers are completely dry). Drinking a cup of chamomile tea before bed can help you relax and have a more restful night's sleep.

Chamomile health benefits:

- Improves overall skin health
- Relieves pain
- Aids sleep
- Reduces inflammation and swelling
- Rich source of antioxidants
- Relieves congestion

Common uses:

- Brew dried flowers for a tea
- Inhale essential oil
- Apply essential oils topically

05. DANDELION

5. Dandelion

You should think twice before removing those pesky dandelions (*taraxacum*) from your front yard! Dandelions are not only edible, but they are also full of health benefits. These medicinal plants are packed with things that are great for you: vitamin K, vitamin C, iron, calcium and more. These vitamins and minerals help support strong bone and liver health. All parts of a dandelion are useful and good for you. For example, dandelion roots are commonly used for teas, the leaves are used as garnishes for different dishes and dandelion sap is great for your skin!

Dandelion health benefits:

- Detoxifies liver and supports overall liver health
- Treats skin infection

- Supports overall bone health
- Treats and helps prevent urinary infections

Common uses:

- Brew roots for a tea
- Use leaves as garnish for dishes
- Consume as a pill

06. ECHINACEA

6. Echinacea

Echinacea (*echinacea purpurea*) is also commonly known as purple coneflower. This is another flower that is normally used in tea to help soothe different symptoms and to strengthen the immune system. This popular herb is used most often to accelerate recovery from the common cold. It's important to note that echinacea can cause negative effects like nausea and dizziness if taken consistently in large doses.

Echinacea health benefits:

- Treats and helps prevent urinary tract infections
- Strengthens immune system
- Relieves upper respiratory issues
- Fights infections
- Alleviates symptoms from the common cold

Common uses:

- Brew roots, leaves and flowers for a tea
- Consume as a pill

07. GARLIC

7. Garlic

Garlic (*allium sativum*) helps keep away vampires and unwanted diseases! This super plant is great for fighting infections, aiding with cholesterol management and much more. Eating garlic on a regular basis is good for your overall health and easy to incorporate into a wide array of dishes. Raw garlic is the most potent, so try eating it uncooked for the most health benefit.

Garlic health benefits:

- Helps prevent heart disease
- Lowers cholesterol and blood pressure
- Prevents dementia, Alzheimer's and similar degenerative diseases
- Improves digestive health

Common uses:

- Use as ingredient or garnish for dishes
- Consume raw

08. LAVENDER

8. Lavender

Lavender (*lavandula*) is popular for its soothing scent and ability to calm the nerves. Lavender tea is another drink you can whip up to help you unwind after a long day and have a good night's rest. Lavender oil is also popular for massage treatments, aromatherapy and even hair treatment!

Lavender health benefits:

- Eases tension and reduces stress
- Relieves headaches and migraines
- Aids sleep
- Supports healthy hair and skin
- Fights acne
- Relieves pain
- Treats respiratory problems

Common uses:

- Brew flowers for a tea
- Use essential oil in a diffuser
- Apply essential oil topically

09. LEMON BALM

9. Lemon Balm

Lemon balm (*melissa officinalis*) is a longstanding medicinal plant used to help relieve stress and ward off insects! An intense amount of stress can cause complications for many functions of the body, so minimal stress is ideal for a healthy functioning body. This lemony plant is delicious and easily used in several dishes like teas, ice cream and more. Many people consume lemon balm tea to help relieve anxiety, stress and even to calm restless kids.

Lemon balm benefits:

- Calms restlessness, anxiety and stress
- Reduces inflammation
- Treats cold sores
- Soothes menstrual cramps

Common uses:

- Brew leaves for a tea
- Garnish for dishes and desserts
- Apply tea or essential oil topically

10. MARIGOLD

10. Marigold

Marigolds (*tagetes*) are fragrant plants that many turn to in order to improve their overall skin health. These vibrant flowers carry a lot of antioxidants and other healthy compounds that make them the perfect choice to keep in your home! These plants not only keep your body healthy, but also help keep insects away.

Marigold health benefits:

- Soothes skin and treats skin diseases
- Reduces inflammation
- Strong antibacterial and antiseptic properties
- Treats ear pain and infection
- Strengthens eyes

Common uses:

- Brew dried flowers for a tea
- Apply essential oil or cream topically
- Sprinkle as a garnish for dishes

11. PARSLEY

11. Parsley

Parsley (*petroselinum crispum*) is a delicious garnish that's helpful for supporting your immune system, bone health and digestive health. The high concentration of antioxidants, vitamin K and other compounds help make this plant an all-around powerhouse herb for your body. Parsley is also a good herb to reach for if you're suffering from halitosis, also known as bad breath!

Parsley health benefits:

- Relieves bloating and supports digestive health
- Fights bad breath
- Supports bone health
- Rich source of antioxidants

Common uses:

- Sprinkle as a garnish for dishes
- Create a juice or brew for a tea

12. PEPPERMINT

12. Peppermint

Peppermint (*mentha × piperita*) is a fresh herb that we taste in gum, toothpaste and desserts. This herb makes a tasty tea and helps relieve tummy aches, nausea and muscle pain (just to name a few). Peppermint tea is a good choice for pregnant moms who suffer from occasional morning sickness.

Peppermint health benefits:

- Relieves allergies
- Soothes muscle pain
- Relieves headaches
- Reduces nausea, gas and indigestion
- Supports digestive health
- Treats bad breath
- Highly antibacterial

Common uses:

- Brew leaves for a tea
- Apply essential oil topically
- Inhale essential oil

13. ROSEMARY

13. Rosemary

Rosemary (*rosmarinus officinalis*) is full of vitamins and minerals that help support many different functions in the body. For instance, rosemary is great for improving memory and also supports hair growth. This means a cup of rosemary tea is great for anyone heading into a night of studying or a person fighting a receding hairline!

Rosemary health benefits:

- Reduces inflammation
- Improves blood circulation
- Improves memory and enhances overall brain function
- Treats bad breath
- Supports liver health
- Supports hair growth

Common uses:

- Brew dried leaves for a tea
- Sprinkle as a garnish for dishes
- Apply essential oil topically

14. SAGE

14. Sage

Sage (*salvia officinalis*) is another medical plant that helps support memory and combat degenerative diseases. Sage is also well-known for managing diabetes with its ability to naturally lower glucose levels. This plant is a popular ingredient for several dishes and beauty products, so you can easily reap the benefits of sage in a multitude of ways!

Sage health benefits:

- Improves memory and enhances overall brain function
- Supports digestive health
- Strengthens immune system
- Treats and helps manage diabetes
- Rich in antioxidants
- Improves skin health

Common uses:

- Brew fresh leaves for tea
- Sprinkle as a garnish for dishes
- Inhale essential oil
- Apply essential oil topically

15. ST. JOHN'S WORT

15. St. John's Wort

St. John's Wort (*hypericum perforatum*) is primarily known as a natural way to relieve symptoms of depression. It's used to treat anxiety, mood swings, feelings of withdrawal and symptoms of obsessive-compulsive disorder. These medicinal plants are usually consumed as a concentrated pill or applied topically as an ointment. It's important to note that St. John's Wort can interact with a number of medications, so (as with all plants on this list) consult your doctor before consuming or applying this plant to your body.

St. John's Wort health benefits:

- Helps relieve symptoms of depression
- Relieves anxiety and helps manage mood
- Reduces inflammation
- Soothes skin irritation

Common uses:

- Consume as a pill
- Brew fresh flowers for tea
- Apply topically as essential oil or ointment

16. THYME

16. Thyme

Thyme (*thymus vulgaris*) is a popular herb used in cooking. Thymol is found in thyme and is commonly found in mouthwash and vapor rubs. This compound gives thyme its strong antifungal and antibacterial properties. Thyme's antifungal properties also helps prevent food borne illnesses since it can decontaminate food and prevent infections in the body.

Thyme health benefits:

- Soothes sore throats and coughs
- Improves blood circulation
- Treats respiratory problems
- Supports immune system

Common uses:

- Sprinkle as a garnish for meals
- Brew fresh leaves for tea
- Apply topically as a cream

Next time you have a pesky pain or symptom, try reaching for one of these medicinal plants!

Medicinal Plants to Keep in Your House Germany Edition

Heilpflanzen sind nützlich, um häufig auftretende Beschwerden zu behandeln. Sie können nach bestimmten Heilpflanzen greifen, um Kopfschmerzen, Bauchschmerzen und sogar Irritationen durch Insektenstiche zu lindern. Pflanzen können in Teesorten konsumiert, als Beilage verwendet, als ätherisches Öl topisch angewendet oder als Pille konsumiert werden.

Es ist wichtig zu bedenken, dass Sie immer Ihren Arzt konsultieren sollten, bevor Sie etwas Neues für Ihren Körper konsumieren oder verwenden. Wenn Sie sich für den Anbau einiger dieser Pflanzen entscheiden, beachten Sie die Pflegerichtlinien der Pflanze und verwenden Sie keine Pestizide oder andere schädliche Chemikalien für Ihre Pflanzen. Sie wollen keine dieser Chemikalien in oder auf Ihrem Körper!

Um Ihnen bei der Entscheidung zu helfen, welche Pflanzen für Sie am besten geeignet sind, haben wir unsere wichtigsten Heilpflanzen, ihre bemerkenswerten gesundheitlichen Vorteile und ihre Verwendung zusammengefasst.

01. BASIL

1. Basilikum

Basilikum (*ocimum basilicum*) ist ein weit verbreitetes Kraut, das zum Garnieren von Salaten, Nudeln und vielen anderen Gerichten verwendet wird, um einen köstlichen Geschmack hinzuzufügen. Dank der im Basilikum enthaltenen Vitamine und Mineralien wie Vitamin K und Eisen hilft dieses Kraut bei der Bekämpfung von Volkskrankheiten. Zum Beispiel hilft das Mangan im Basilikum, verschiedene Verbindungen in Ihrem Körper zu metabolisieren. Heiliges Basilikum, allgemein als Tulsi bezeichnet, ist eine spezielle Basilikumart, die aus Indien stammt. Es gilt als heilige Pflanze, die in Tees, Salben und vielem mehr verwendet wird, um eine Vielzahl von Krankheiten wie Fieber und Diabetes zu behandeln. Diese Art hat einen viel stärkeren Geschmack als gewöhnliches Basilikum!

Nutzen für die Gesundheit des Basilikums:

- Reduziert Stress
- Starke antibakterielle Eigenschaften
- Reichhaltige Quelle von Antioxidantien
- Verhindert einige schädliche Auswirkungen des Alterns
- Reduziert Entzündungen und Schwellungen
- Stärkt Knochen und Leber
- Steigert die Immunität
- Steigert den Stoffwechsel
- Verbessert die Verdauung

Allgemeine Verwendungen:

- Streuen Sie als Beilage für Geschirr
- Als Zutat für Smoothies verwenden

02. CATNIP

2. Katzenminze

Katzenminze (*Nepeta Cataria*) ist eine lustige Pflanze für Katzen. Die meisten Katzen fühlen sich von der Pflanze angezogen und rollen in ihrer Nähe herum, da ihr Aroma stimulierend wirkt. Diese Heilpflanzen wirken auch als Beruhigungsmittel für Katzen, wenn sie verzehrt werden. Für den Menschen hingegen wird es normalerweise als Stressabbau, Schlafmittel und Lösung für Hautprobleme eingesetzt. Der Großteil seiner gesundheitlichen Vorteile beruht auf dem Vorhandensein von Nepetalacton, Thymol und anderen Verbindungen, die diese Pflanze für Sie und Ihren pelzigen Freund großartig machen.

Nutzen für die Gesundheit der Katzenminze:

- Weist Insekten ab und lindert Irritationen durch Insektenstiche
- Beruhigt Unruhe, Angst und Stress

- Lindert Magenbeschwerden
- Beschleunigt die Genesung von Erkältungen und Fieber

Allgemeine Verwendungen:

- Brauen Sie Blätter für einen Tee
- Trocknen Sie die Blätter und verbrennen Sie sie, um das Aroma freizusetzen
- Tragen Sie ätherische Öle oder Blätter topisch auf

03. CAYENNE PEPPER

3. Cayenne-Pfeffer

Cayenne-Pfeffer (*Capsicum Annuum*) verleiht jeder Mahlzeit oder jedem Getränk einen würzigen Kick und ist bei vielen Menschen ein beliebtes Entgiftungsmittel. Capsaicin ist die Verbindung, die für die Würzigkeit von Cayenne verantwortlich ist, aber es ist auch für einige seiner gesundheitlichen Vorteile verantwortlich. Einige dieser Vorteile sind Schmerzlinderung und niedriger Cholesterinspiegel.

Gesundheitliche Vorteile von Cayennepfeffer:

- Entgiftet den Körper
- Steigert den Stoffwechsel
- Erleichtert Magenverstimmung und fördert die Verdauung
- Verbessert die Durchblutung
- Lindert Schmerzen

Allgemeine Verwendungen:

- Fügen Sie Saucen, Gewürzmischungen, Dressing und anderen Gerichten hinzu
- Als Pille einnehmen

04. CHAMOMILE

4. Kamille

Kamille (*Matricaria chamomilla*) enthält eine hohe Konzentration an Antioxidantien, die es zu einer großartigen Pflanze zur Linderung einer Vielzahl von Krankheiten macht. Kamille wird normalerweise als Tee konsumiert und Sie können sich Ihre eigenen zu Hause zubereiten, indem Sie getrocknete Kamillenblüten brauen (stellen Sie nur sicher, dass die Blüten vollständig trocken sind). Wenn Sie vor dem Schlafengehen eine Tasse Kamillentee trinken, können Sie sich entspannen und erholsamer schlafen.

Nutzen für die Gesundheit der Kamille:

- Verbessert die allgemeine Hautgesundheit
- Lindert Schmerzen
- Hilft beim Schlafen
- Reduziert Entzündungen und Schwellungen

- Reichhaltige Quelle von Antioxidantien
- Lindert Staus

Allgemeine Verwendungen:

- Brauen Sie getrocknete Blumen für einen Tee
- Ätherisches Öl einatmen
- Tragen Sie ätherische Öle topisch auf

05. DANDELION

5. Löwenzahn

Sie sollten zweimal überlegen, bevor Sie diesen lästigen Löwenzahn (*Taraxacum*) aus Ihrem Vorgarten entfernen! Löwenzahn ist nicht nur essbar, sondern auch voller gesundheitlicher Vorteile. Diese Heilpflanzen sind vollgepackt mit Dingen, die gut für Sie sind: Vitamin K, Vitamin C, Eisen, Kalzium und mehr. Diese Vitamine und Mineralien unterstützen die Gesundheit von Knochen und Leber. Alle Teile eines Löwenzahns sind nützlich und gut für Sie. Zum Beispiel werden Löwenzahnwurzeln häufig für Tees verwendet, die Blätter werden als Beilage für verschiedene Gerichte verwendet und Löwenzahnsaft ist großartig für Ihre Haut!

Löwenzahn Nutzen für die Gesundheit:

- Entgiftet die Leber und unterstützt die allgemeine Gesundheit der Leber
- Behandelt Hautinfektionen

- Behandelt und beugt Harnwegsinfekten vor
- Unterstützt die allgemeine Knochengesundheit

Allgemeine Verwendungen:

- Brauen Sie Wurzeln für einen Tee
- Verwenden Sie Blätter als Beilage für Gerichte
- Als Pille einnehmen

06. ECHINACEA

6. Echinacea

Echinacea (*Echinacea purpurea*) ist auch als purpurroter Sonnenhut bekannt. Dies ist eine weitere Blume, die normalerweise im Tee verwendet wird, um verschiedene Symptome zu lindern und das Immunsystem zu stärken. Dieses beliebte Kraut wird am häufigsten verwendet, um die Erholung von der Erkältung zu beschleunigen. Es ist wichtig zu beachten, dass Echinacea negative Wirkungen wie Übelkeit und Schwindel hervorrufen kann, wenn es regelmäßig in großen Dosen eingenommen wird.

Gesundheitliche Vorteile von Echinacea:

- Behandelt und beugt Infektionen der Harnwege vor
- Stärkt das Immunsystem
- Lindert Probleme der oberen Atemwege
- Bekämpft Infektionen

- Lindert Erkältungssymptome

Allgemeine Verwendungen:

- Brauen Sie Wurzeln, Blätter und Blüten für einen Tee
- Als Pille einnehmen

07. GARLIC

7. Knoblauch

Knoblauch (*allium sativum*) hilft, Vampire und unerwünschte Krankheiten fernzuhalten! Diese Superpflanze ist ideal zur Bekämpfung von Infektionen, zur Unterstützung des Cholesterinmanagements und für vieles mehr. Regelmäßiger Verzehr von Knoblauch ist gut für Ihre allgemeine Gesundheit und lässt sich leicht in eine Vielzahl von Gerichten integrieren. Roher Knoblauch ist am wirkungsvollsten. Versuchen Sie daher, ihn ungekocht zu essen, um die Gesundheit zu fördern.

Nutzen für die Gesundheit des Knoblauchs:

- Hilft bei der Vorbeugung von Herzerkrankungen
- Senkt Cholesterin und Blutdruck
-

- Verhindert Demenz, Alzheimer und ähnliche degenerative Erkrankungen
- Verbessert die Verdauungsgesundheit

Allgemeine Verwendungen:

- Verwenden Sie als Zutat oder Garnitur für Gerichte
- Roh verzehren

08. LAVENDER

8. Lavendel

Lavendel (*Lavandula*) ist beliebt für seinen beruhigenden Duft und seine Fähigkeit, die Nerven zu beruhigen. Lavendeltee ist ein weiteres Getränk, mit dem Sie nach einem langen Tag die Seele baumeln lassen und eine gute Nachtruhe genießen können. Lavendelöl ist auch für Massageanwendungen, Aromatherapie und sogar Haarbehandlung beliebt!

Lavendel Nutzen für die Gesundheit:

- Lindert Verspannungen und baut Stress ab
- Lindert Kopfschmerzen und Migräne
- Hilft beim Schlafen
- Unterstützt gesundes Haar und Haut
- Bekämpft Akne
-

- Lindert Schmerzen
- Behandelt Atemprobleme

Allgemeine Verwendungen:

- Brauen Sie Blumen für einen Tee
- Verwenden Sie ätherisches Öl in einem Diffusor
- Tragen Sie ätherisches Öl topisch auf

09. LEMON BALM

9. Zitronenmelisse

Zitronenmelisse (*Melissa Officinalis*) ist eine langjährige Heilpflanze, die verwendet wird, um Stress abzubauen und Insekten abzuwehren! Starker Stress kann zu Komplikationen bei vielen Körperfunktionen führen, daher ist minimaler Stress ideal für einen gesund funktionierenden Körper. Diese Zitronenpflanze ist köstlich und kann leicht in verschiedenen Gerichten wie Tees, Eis und vielem mehr verwendet werden. Viele Menschen trinken Zitronenmelissentee, um Angstzustände und Stress abzubauen und um unruhige Kinder zu beruhigen.

Zitronenmelisse Vorteile:

- Beruhigt Unruhe, Angst und Stress
- Reduziert die Entzündung

- Behandelt Fieberbläschen
- Beruhigt Menstruationsbeschwerden

Allgemeine Verwendungen:

- Brauen Sie Blätter für einen Tee
- Für Gerichte und Desserts garnieren
- Tee oder ätherisches Öl topisch auftragen

10. MARIGOLD

10. Ringelblume

Ringelblumen (*Tagetes*) sind duftende Pflanzen, an die sich viele wenden, um ihre allgemeine Hautgesundheit zu verbessern. Diese lebendigen Blumen enthalten viele Antioxidantien und andere gesunde Wirkstoffe, die sie zur perfekten Wahl für die Aufbewahrung in Ihrem Zuhause machen! Diese Pflanzen halten nicht nur Ihren Körper gesund, sondern halten auch Insekten fern.

Vorteile für die Gesundheit von Ringelblumen:

- Beruhigt die Haut und behandelt Hautkrankheiten
- Reduziert die Entzündung
- Starke antibakterielle und antiseptische Eigenschaften
- Behandelt Ohrenschmerzen und Infektionen
- Stärkt die Augen

Allgemeine Verwendungen:

- Brauen Sie getrocknete Blumen für einen Tee
- Tragen Sie ätherisches Öl oder Creme topisch auf
- Streuen Sie als Beilage für Geschirr

11. PARSLEY

11. Petersilie

Petersilie (*Petroselinum crispum*) ist eine köstliche Beilage, die Ihr Immunsystem, Ihre Knochengesundheit und Ihre Verdauungsgesundheit unterstützt. Die hohe Konzentration an Antioxidantien, Vitamin K und anderen Verbindungen macht diese Pflanze zu einem Allround-Kraftkraut für Ihren Körper. Petersilie ist auch ein gutes Kraut, wenn Sie an Mundgeruch leiden, auch bekannt als Mundgeruch!

Vorteile für die Gesundheit von Petersilie:

- Lindert Blähungen und fördert die Verdauung
- Bekämpft Mundgeruch
- Unterstützt die Knochengesundheit
- Reichhaltige Quelle von Antioxidantien

Allgemeine Verwendungen:

- Streuen Sie als Beilage für Geschirr
- Stellen Sie einen Saft her oder brauen Sie für einen Tee

12. PEPPERMINT

12. Pfefferminze

Pfefferminze (*Mentha × Piperita*) ist ein frisches Kraut, das wir in Gummi, Zahnpasta und Desserts schmecken. Dieses Kraut macht einen leckeren Tee und lindert Bauchschmerzen, Übelkeit und Muskelschmerzen (um nur einige zu nennen). Pfefferminztee ist eine gute Wahl für schwangere Mütter, die gelegentlich unter morgendlicher Übelkeit leiden.

Nutzen für die Gesundheit von Pfefferminze:

- Lindert Allergien
- Lindert Muskelschmerzen
- Lindert Kopfschmerzen
- Reduziert Übelkeit, Gas und Verdauungsstörungen
- Unterstützt die Gesundheit des Verdauungssystems

- Behandelt Mundgeruch
- Sehr antibakteriell

Allgemeine Verwendungen:

- Brauen Sie Blätter für einen Tee
- Tragen Sie ätherisches Öl topisch auf
- Ätherisches Öl einatmen

13. ROSEMARY

13. Rosmarin

Rosmarin (*Rosmarinus officinalis*) ist voller Vitamine und Mineralien, die viele verschiedene Funktionen im Körper unterstützen. Zum Beispiel ist Rosmarin großartig für die Verbesserung des Gedächtnisses und unterstützt auch das Haarwachstum. Dies bedeutet, dass eine Tasse Rosmarin-Tee ideal für jeden ist, der in eine Nacht des Studierens eintritt oder gegen einen zurückweichenden Haaransatz kämpft!

Nutzen für die Gesundheit von Rosmarin:

- Reduziert die Entzündung
- Verbessert die Durchblutung
- Verbessert das Gedächtnis und verbessert die allgemeine Gehirnfunktion
- Behandelt Mundgeruch

- Unterstützt die Gesundheit der Leber
- Unterstützt das Haarwachstum

Allgemeine Verwendungen:

- Brauen Sie getrocknete Blätter für einen Tee
- Streuen Sie als Beilage für Geschirr
- Tragen Sie ätherisches Öl topisch auf

14. SAGE

14. Salbei

Salbei (*salvia officinalis*) ist eine weitere Heilpflanze, die das Gedächtnis unterstützt und degenerative Krankheiten bekämpft. Salbei ist auch bekannt für die Behandlung von Diabetes mit seiner Fähigkeit, den Glukosespiegel auf natürliche Weise zu senken. Diese Pflanze ist eine beliebte Zutat für verschiedene Gerichte und Schönheitsprodukte, sodass Sie die Vorteile von Salbei auf vielfältige Weise nutzen können!

Nutzen für die Gesundheit von Salbei:

- Verbessert das Gedächtnis und verbessert die allgemeine Gehirnfunktion
- Unterstützt die Gesundheit des Verdauungssystems
- Stärkt das Immunsystem

- Behandelt und hilft bei der Behandlung von Diabetes
- Reich an Antioxidantien
- Verbessert die Gesundheit der Haut

Allgemeine Verwendungen:

- Zum Tee frische Blätter aufbrühen
- Streuen Sie als Beilage für Geschirr
- Ätherisches Öl einatmen
- Tragen Sie ätherisches Öl topisch auf

15. ST. JOHN'S WORT

15. Johanniskraut

Johanniskraut (hypericum perforatum) ist in erster Linie als natürlicher Weg zur Linderung von Depressionssymptomen bekannt. Es wird zur Behandlung von Angstzuständen, Stimmungsschwankungen, Rückzugsgefühlen und Symptomen von Zwangsstörungen angewendet. Diese Heilpflanzen werden in der Regel als konzentrierte Pille verzehrt oder topisch als Salbe angewendet. Es ist wichtig zu beachten, dass Johanniskraut mit einer Reihe von Medikamenten interagieren kann. Konsultieren Sie daher (wie bei allen Pflanzen auf dieser Liste) Ihren Arzt, bevor Sie diese Pflanze verzehren oder auf Ihren Körper auftragen.

Die gesundheitlichen Vorteile von Johanniskraut:

- Hilft bei Depressionssymptomen
- Lindert Angstzustände und hilft beim Umgang mit Stimmung
- Reduziert die Entzündung
- Lindert Hautreizungen

Allgemeine Verwendungen:

- Als Pille einnehmen
- Brauen Sie frische Blumen zum Tee
- Topisch als ätherisches Öl oder Salbe auftragen

16. THYME

16. Thymian

Thymian (*Thymus vulgaris*) ist ein beliebtes Kraut, das beim Kochen verwendet wird. Thymol ist in Thymian enthalten und wird häufig in Mundwasser und Dampfreiben verwendet. Diese Verbindung verleiht Thymian seine starken antimykotischen und antibakteriellen Eigenschaften. Die antimykotischen Eigenschaften von Thymian tragen auch zur Vorbeugung von durch Lebensmittel übertragenen Krankheiten bei, da sie Lebensmittel dekontaminieren und Infektionen im Körper verhindern können.

Nutzen für die Gesundheit von Thymian:

- Beruhigt Halsschmerzen und Husten
- Verbessert die Durchblutung
- Behandelt Atemprobleme
- Unterstützt das Immunsystem

Allgemeine Verwendungen:

- Zum Garnieren zu den Mahlzeiten einstreuen
- Zum Tee frische Blätter aufbrühen
- Topisch als Creme auftragen

Wenn Sie das nächste Mal lästige Schmerzen oder Symptome haben, greifen Sie nach einer dieser Heilpflanzen!

References

Collins, Minta (2000). Medieval Herbals: The Illustrative Traditions. University of Toronto Press. ISBN 978-0-8020-8313-5.

Sumner, Judith (2000). The Natural History of Medicinal Plants. Timber Press. p. 17. ISBN 978-0-88192-483-1.

Grene, Marjorie (2004). The philosophy of biology: an episodic history. Cambridge University Press. p. 11. ISBN 978-0-521-64380-1.

Jacquart, Danielle (2008). "Islamic Pharmacology in the Middle Ages: Theories and Substances". European Review.

Singh, Amritpal (2016). Regulatory and Pharmacological Basis of Ayurvedic Formulations. CRC Press. pp. 4–5. ISBN 978-1-4987-5096-7.

Newall, Carol A.; et al. (1996). Herbal medicines: a guide for health-care professionals. Pharmaceutical Press. ISBN 978-0-85369-289-8.

Lightning Source UK Ltd.
Milton Keynes UK
UKHW020412070220
358284UK00002B/12